HOW TO TEACH SCIENCE THROUGH INVESTIGATIONS

Getting students to learn science by having them do science.

By: Christopher P. Garside
Seven Sides Publishing

Seven Sides Publishing of Cypress, TX has a mission to improve teaching and the understanding of Science. To contact us, please send an email to:

simpleinvestigations@sevensidespublishing.com

Copyright © 2018 and 2021 by Christopher P. Garside

All rights reserved. No part of this publication can get reproduced, stored in a retrieval system, or transmitted in any form or by any means, electronic, mechanical, photocopying, recording, or otherwise, without the prior written permission of Christopher P. Garside and Seven Sides Publishing.

ISBN: 9781976931192

Published by: *Seven Sides Publishing*, Cypress, TX.

Made in the United States of America

Pictures and illustrations are from *The Big Box of Art* (2004), produced by Hemera Technologies Inc., Gatineau, Quebec, Canada.

Table of Contents

Introduction — **page 3**

Chapter 1
What is Wrong? — **page 7**

Chapter 2
Why Teach Through Investigations? — **page 17**

Chapter 3
What Are Investigations? — **page 31**

Chapter 4
How to Teach Through Investigations — **page 37**

Chapter 5
Systemic Challenges You May Face — **page 59**

References — **page 74**

Introduction

Dedication

I would like to dedicate this book to all the teachers who tirelessly work hard every day. They do not work for a paycheck but to improve student's lives. Helping them find and develop who they are and what they can become. To all the teachers who endlessly try to improve their craft as Educators in the most underpaid, underappreciated career. I dedicate this to you, the unselfish, who spend their extra time and their own money to help provide students the best educational and personal services schools and districts could not be able to provide without your extra efforts and the love you put into your students. I want to give a particular dedication to the departed Daniel Felske, who passed away from cancer many years ago. Dan inspired me down this path of focusing on the science process skills a long time ago while we were both still working in H.I.S.D., I at Scarborough High School and Eastwood Academy, and he at Booker T. Washington. He was the one who showed me that students first need to develop science literacy skills before they can learn Biology to do well on the first generation of Biology E.O.C. statewide tests for Texas. We battled back and forth for bragging rights for the highest passing percentages for years before he left the classroom for an administrative position shortly before his death.

Acknowledgments

I want to thank my friends Chris Hecker, John Henderson, Tina King, and Walter Pitlock for reviewing the transcript for this book. Thank you for your long hours of effort to help me help science teachers through this book. Your feedback has been constructive and valuable to me in producing this final product.

About the Book

This book will show you what is wrong with the way we typically educate students in science today, why we need to change, what society/business needs in workers of the 21st century, and how we as science teachers can meet the needs of the students. This book will point out systemic challenges you may face and other strategies you can use to help you implement the process of *Teaching Science Through Investigations*. It will define and categorize the types of investigations that can be done. It will then show you a more efficient way to educate students, so they can learn science by learning the science content and skills together at the same time. Having students learn through investigations will help them understand the vocabulary in the context of other science vocabulary and content. This strategy allows students to chunk information more quickly so it can be learned more efficiently. This information is not new but is presented in a way that is easy and unintimidating to implement. This strategy has been called inquiry learning and investigative learning in the past. Implementing this process has been a struggle for teachers because of 2 things. First, it takes too much time to develop and find enough good labs and equipment to go with a class we are teaching. Second, we think we need to give lecture notes and PowerPoints, but the students would be better off without them. Teachers are addicted to giving lecture notes to students because this was the primary way we learned information growing up. This book will show you how and why lecturing is the most inefficient strategy to pass on information to students. It is a waste of so much instructional time that we don't get a chance to practice the science process skills, problem-solving skills, and higher application of much of the science content that our students could apply to all aspects of their lives and prepare them for prospective careers. We do not have to rob them of these skills any longer. This book will show you how to prepare our students for the 21st century by having students do science to learn science in your school/classroom.

Chapter 1

What is Wrong?

When I was a sophomore in high school, I took a Chemistry class. Because of what I had experienced in that class, I avoided science like the plague until my second year of college. As a kid, I loved everything about science. I did not miss an episode of *Cosmos* with Carl Sagan and looked forward to a new episode each week. I had lots of science-related toys, and I loved science fiction movies. I even wanted to be an astronomer when I grew up. I loved anything that had to do with science up until this chemistry class. What I experienced in that chemistry class made me hate science, and I wanted nothing to do with learning any other science in the future.

I saw no point in it because of the way it was presented to me. I saw no connection to real life. I did not get to see the concepts in action; we only talked about them. I saw nothing but many disconnected facts that I had to memorize with lots of busy worksheets and vocabulary to learn. I did not understand why I needed to know any of it; I saw no relevance. I could count how many labs I did in my chemistry class on one hand and have three fingers still left over. It was not a challenging class, but it was boring and filled with busy work. I was not taught how to think because most of the course was about taking lecture notes and memorizing those notes and the periodic table. I was not required to take any more science classes to graduate high school, so I did not until I was forced to take one in college. I was a Math Education Major and was required to take a science class, so I had to decide which one I would take. Since I had already taken Biology in middle school and did well in it, I chose to take Biology. In that class, I finally got to do some labs.

I got to see how information could be observed and that I could learn independently of a teacher. This class related science to real life for me, and I thought Biology was the coolest thing in the world. At that time, I took second-semester Calculus, and it was kicking my butt, so I started looking for a new major. I remember one lab in the Biology class where I was looking through a binocular microscope. I saw the beating heart of a baby chicken inside of an open egg. I could not believe what I was looking at; I was hooked, and I found my new major and a new love: Biology Education. It was not until I started teaching that I also found a passion for both Chemistry and Physics.

To get my first job teaching Biology, I also needed to teach a Physical Science class where I ended up learning about fundamental Physics and Chemistry, both of which I took in college but hated because they were taught with lots of lectures and math, which were way over my head. When I taught Physical Science, I found there were so many engaging experiments that helped me effortlessly learn content that I could do with the students that made me say, "Wow, where was all this when I was learning in school? Why did my teachers not do these things with me?" I do not remember much from my high school Chemistry and very little from my college Chemistry and Physics classes. I remembered a lot from the Physical Science classes I taught because of the experiences I created through labs. I remembered so much of what I taught myself through preparing the labs and virtual labs I used while teaching my students. We remember about 5-20% of what we hear, about 75-80% of what we experience, and about 90-95% of what we teach. From that information alone, teachers should be spending most of the class time providing opportunities where students will get to experience science and teach it to their fellow students somehow. Because of the experience I created in the classroom, and the fact that I sought out why the concepts were essential and applicable to real life, it gave the relevance of the concepts not only to me but my students, which put the information in my long term memory without having to do very much work.

How many students go through our schools that would love science if they just got to see it, experience it, and find out how it is used every day in our lives? Many adults have no concept of how science works because they were never taught how to do it in school. They were taught that it is just a bunch of random facts that have nothing to do with each other like I was taught while in high school. This is how I think most people view science, just a bunch of unorganized facts that have no relation to anything in real life. They do not understand that it is a process, a way of thinking, and a way to discover the world. Many people do not understand how to observe the world or gather information to find out how it works. Not having those skills is why so many people pay workers to install, fix, and maintain appliances in different parts of their homes that could quickly be done themselves. For example, when your air conditioner goes out, you call some big-name air conditioning company. They come out and say you need a whole new system that costs $20,000 when it is just a burned-out capacitor ($28), a valve cap ($.50), or you need a new motherboard (about $150). Solving this kind of problem could be done for anything in our homes, like maintaining appliances like washers, a clothes drier, pool pumps, or plumbing repairs. Just about anything you need to do to maintain your home could be fixed and repaired by you if you develop the science process skills to observe, problem-solve, and know how to learn new things. If we had been more science-savvy, we could save our families time and money if we were taught how to think like scientists.

Many jobs require people to have a skill set that should have been taught and developed in science classes. We need a workforce skilled in problem-solving where they can solve problems that do not exist yet. So many of these jobs go unfilled because so few people go into science-related fields to get educated with the knowledge and skills they could use for these jobs. There are so few people educated in the United States who are qualified for these jobs in the U.S. that companies have to recruit people from outside the U.S. to fill them with qualified individuals. Not having eligible Americans is happening because of the way science content is commonly taught to middle and high school students. Teachers don't usually have students practice the skills needed to acquire knowledge of this world and make sense of that knowledge, so it can be used to solve real problems in the future. Not having students practice skills may be why so few people have common sense (that is why common sense is not very common).

Isn't that what science is supposed to teach? We are to teach skills to learn about the world around us independent of a textbook. Too many teachers use the textbook as a bible for their curriculum instead of a resource.

I've found that the best use of our textbooks is as a way to prop up my ramps for acceleration experiments in my Physics class. Now, with virtual textbooks, I can't even use them for that. Districts and states do not want to give us hard copies because it costs more money to have a real physical book than a virtual one. How much money do you think we could save by not having any science textbooks at all? We would not have to renew textbook adoptions every few years. So we must ask ourselves, do we really need textbooks? Or better yet, should we be using them? Students can learn everything in science through real-life experiments, virtual labs, and research on the internet. There could be a lot of money saved by school districts from not buying textbooks, and that money could be spent somewhere else in the budget to buy more equipment, improve technology, or raise teachers' salaries. We do not need textbooks. All the information students need is provided in real life and on the search engines of the internet. All the students need are the skills to find and use this information. Having textbooks around might be something holding our students back from performing more investigations.

How does your brain remember information best? Do you learn best from someone telling you information or by experiencing it and searching for the information yourself? Why does most job training require people to learn what to do in their jobs by participating in their job as they do it? Or why do people have to follow or shadow someone before they start working the job on their own? Some will ask, "Isn't reading about how to do your job or having someone tell you what to do enough?" Real-life shows us it is not. Businesses invest money in their people, and they know that the best way to prepare them to learn a new job is through experiencing it, not just reading about it. All students from Special Education to Gifted and Talented learn best by doing.

Observing the world around you, then manipulating variables until you figure out what you need is a thinking process and skill that has to be developed through lots of practice. Traditional education has failed us because teachers do not teach science as a process in class. Most of the science process is taught through lecture notes in the first unit of the year, and the kids never see it again. Many times it is taught more like a History, English, or Math class. Students are put into desks to stay seated in rows, listen to lectures, do vocabulary activities and worksheets. Research has told us for years how students should learn science, but why does it not happen. Some of this has to do with laziness because it requires time and effort to change what you are doing. Trying to teach through investigations is hard because there are few useful resources to help teachers do investigative learning. I noticed big publishing companies have few usable labs. They require too much reading and preparation that no one wants to take the time to read and prepare for them. The teacher's time is precious. They need resources where they can find, modify, or develop investigations efficiently.

Seven Sides Publishing has developed some products to help teachers teach science through investigations with minimal effort. We have Biology, Chemistry, and Physics lab manuals. They are called Simple Investigations. They have been written efficiently where the investigations tell you quickly what materials you will need, the directions, places to report your observations and data, and questions that sum up students' concepts and skills. There are also digital versions available, so you can modify them to tailor them to your needs. They start each section with concept maps that organize vocabulary and concepts that show how each relates to others. Each section ends with a list of websites and virtual investigations that give you even more experiments to complement the manuals' labs. Together there are more than enough investigations that you can do every day and still have some leftover.

These investigations are the ones I had been using and modifying for decades with great success. They are focused on trying to have students learn content and skills at the same time. As I progressed, I saw an unexpected effect. My Special Education and ESL students were having more success. The Special Education and ESL coordinators loved my lessons so much that they loaded me up with their kids. If those kids can do it, anyone can do it. Is change so bad if it will improve student learning? When I later became department chair at a few of the schools I taught, I encouraged the teachers to teach this way. Once teachers saw how it was done and how it benefited the kids, investigations were the norm from then on in their classes. Teachers need to do what we tell our students to do, continue learning and improving our craft for the rest of our careers, and never stop learning until we stop breathing.

I am one of very few teachers that applauded the idea of new standardized End of Course Exams in Texas. I liked it because students would have to learn science process skills to answer these standardized test questions. Most teachers hated it because it meant they had to change what they were doing. They did not want to change to teach to some test, even if that test matched the taught curriculum. When you hear someone say, "Teaching to the test." It is usually in a negative connotation, saying it is a bad thing. Teaching to the test is not a bad thing. Teaching the test is, which students tell me happens all the time in their classes. Students and teachers often tell me that they prepare for the test by practicing a review identical to the test. Teaching the test is flat-out disgraceful, insulting to the teacher and the students, and should never be done unless it is in the context of a high-level take-home test that is quite complicated and requires students to use investigation and synthesis skills on their own. A teacher is supposed to teach the content and skills that a student would need to do well on the test. Good teachers are trained to develop the test first, then the lessons that would allow students to succeed on that test. I do not understand why people talk like this is the wrong thing to do. It is not only the right thing to do; it is the fair thing to do for the students. When science skills and content are in mandated tests, this forces science teachers to have students learn science skills with the content, so they would do well on the exams to make their school look good.

Once those tests were looming on the horizon, I noticed that changes started taking place for the better. There was a push to have teachers change, and everyone was bought into those changes. Once the state took many of those tests away (no EOCs for Chemistry or Physics), teachers in those non-tested subjects backslid to what they were doing before because there was nothing to be accountable for anymore. There was no reason to improve because there was no test to be measured to hold us responsible. So many times, teachers said to me, "There is no End of Course anymore. Why do I need to hold the students to a higher standard? Why put in the effort?" It was easier to do just what they had been doing in the past because they were experienced with those activities, and students would not be tested by the state anymore, so there was no reason to improve their instruction. We need to hold our students to higher standards and develop their skills; so they are college and career ready when they leave us. We want them to be qualified for good jobs available now that will support them and their families. You are not just helping one set of students, but generations could benefit in your students' future that you would have influenced.

What I want to do in the rest of this book is explain why and show how a teacher or a team of teachers can teach where students will learn science by doing science. Thinking like a scientist and learning skills at the same time will carry over to every other aspect of their lives when we teach through investigations. Learning this way will help students better know how to learn and make sense of the world around them, making their lives (in all areas) more manageable and productive in the future.

Chapter 2

Why Teach Through Investigations?

This generation of students is the first generation that does not need us to teach them facts. With the advancement of the internet and search engines, students can find just about anything they want to know on the internet. Anyone can help students learn facts by just putting them in front of a computer or phone and having them answer questions, but students need to apply information. "Good students" tend to be information-rich and analysis poor. They can get and recite bits of information, but they do not know how to apply it in everyday life in new situations. What students need are guides to help them make sense of the information. They must develop skills to learn how this information can explain other concepts and how it can be applied to new situations. They need to have the skills to solve tomorrow's problems that do not exist yet. Teaching through investigations means we will be tapping into Bloom's Taxonomy's highest levels, spending a lot of time in Quadrant D on the Rigor/Relevance Chart. This chart is on page 21 and talked more about later in this chapter. Dr. Willard Dagget is famous for his work with "Quadrant D." He says that the most successful schools live in Quadrant D. They don't just visit it; they live there, beginning to end. Staying in Quadrant D is what we can do when teaching science through investigations. Everyone would benefit greatly by studying his work. It shows us how to be more efficient at running our classrooms, schools, and school districts.

How are we going to teach students those skills if we have students thinking (or not thinking) at the lowest cognitive level at the bottom of Bloom's Taxonomy, i.e., listening to a teacher's lecture, doing a worksheet that goes over that same information, and taking a test that assesses the students at the same low, rote, memorization level? When do students get a chance to practice thinking, observing, analyzing, synthesizing, evaluating, and problem-solving for themselves? Students need to be learning by doing, just like how these top companies train their employees. Why are those companies so successful? Because they know people learn by doing. How are students going to find out how science works if they do not practice it regularly? To have the students get a chance to do this, I used to tell my department we should try to have the students go through the scientific method by having them perform an experiment at least once a unit. For years, I was not able to do that while still giving lecture notes/PowerPoints. I used to think that was a very lofty goal, but now my students can go through that process each day, sometimes more than once, and sometimes even while taking a test. What I will explain will far exceed that once a unit goal and have your students do science every day while seamlessly learning their science concepts.

When interviewing students about the most common method teachers use to teach science, the students tell me most science teachers give lecture notes or PowerPoint presentations, have students read a textbook and answer the questions in the back of the chapter, or do a worksheet while the teacher sits at their desk. These methods usually only make students think at a rote memorization level with no science process, including many higher-order thinking skills learned or practiced. When teachers have been doing something for a long time, they forget how to do anything else, and it becomes a habit. We want to develop new, better practices; and lose the old, bad habits.

If you are familiar with the rigor and relevance chart, a lot of teachers keep their students' brains in quadrant "A" and maybe have some time in quadrant "B" if they are in any quadrant at all (because when a teacher lectures, they cannot prove that a student is even thinking). I am proposing we keep students' brains in quadrants "C" and "D" most of the time. Some might say that you must spend time in quadrant "A" before going to "B" or "C" and then spend time in "B" and "C" before getting to quadrant "D." I have 30 years of experience in teaching science that says you can take students straight to quadrants "C" and "D" right away by learning the concepts and patterns in context through performing investigations; (Dr. Willard Dagget agrees). By understanding the content in the context of other content through an experience, students will recall so much more. Yes, they will have to think differently from what they have been used to in the past. We must think this way when we bring our first baby home from the hospital and have no idea how to keep this thing alive except feed it, clean it, and hold it. I tell my students that the most critical times in their lives require them to learn and problem solve on the run. They might as well practice using their brain this way when not so much is on the line. They need to put effort into their education, not just for themselves but also for those lives that will depend on them later. At some point, they will be leaders of their family, and they need to have skills to guide them through life. Many of my students are already feeling that pressure because they are already parents themselves, and they echo this sentiment for me during class. If you want to know what quadrant you are aiming for with your students, look at your reviews and tests. Are they written where students just have to recall little bits of isolated information or are you having students process information to come up with an answer? If you are having students memorize bits of information, you will have your students in quadrant A. If they

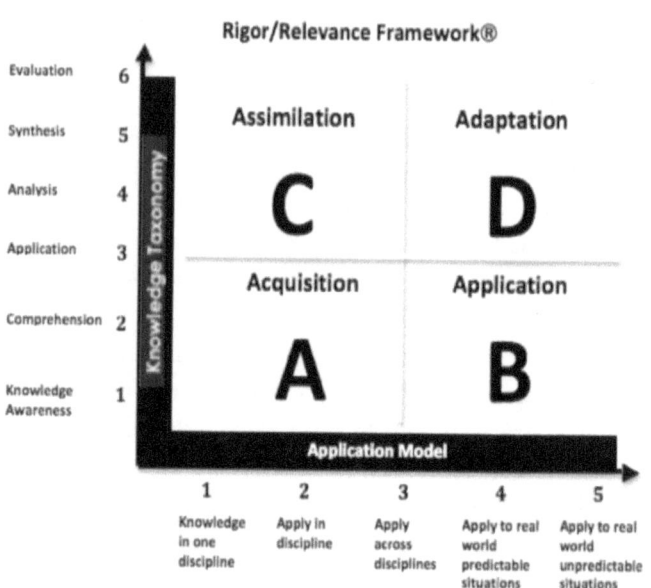

are processing and applying the information to new situations; you will have them in C and D. Look at the chart above and pull out your current activities, reviews, and tests and see which quadrant you have your students spending most of their time in class. Think about where students are on this chart when they are doing investigations. The next chapter defines and organizes investigations for you if you are not sure what investigations are. If you want your students to be ready for a career, you must have them in quadrant D. The majority of State Standards are written for students to be in quadrant C. If you continually have and test your students in quadrant A and B, you are preparing them for menial jobs that they will probably hate and not make very much money and therefore, not provide well for their families. They will have to put more effort into knowing and performing less. I cannot express enough how important it is for us to expose our students to quadrant D as much as we can. If we do not, our students will be missing out on lots of potential opportunities they could achieve if our curriculum is in D, but they will not get the chance if we keep them in A and B.

Research shows that learning a language requires context. The best way to learn a new language; is to be immersed in the culture. When you are learning science vocabulary, much of it is like learning a foreign language. You do not need to define terms when students will do that independently as they research the concepts and develop their own definitions after experiencing the idea. Investigations personalize the knowledge in the context of other concepts without focusing on the teacher's definition by itself. It allows students to learn the concepts as they imprint them in their brains through multiple sensory pathways (using numerous senses), giving them various paths to recall them. An investigation also associates concepts with other concepts showing how they relate to each other in each other's context. Investigating makes it easier for students to then make "chunks" out of the information that allows them to remember so much more efficiently in their long-term memory at higher cognitive levels.

When conducting investigations, students learn how to think like scientists by using their science process skills to discover their content knowledge. What follows is a list of some of these science process skills that are missed when students don't get to do investigations:

- Observing
- Questioning
- Being able to use science equipment: meter sticks, rulers, graduated cylinders, triple beam balances, timing devices, probe ware (like motion detectors, pH meters, temperature probes, dual-range force sensors, oxygen and carbon dioxide probes, and pressure sensors), thermometers, calculators, computers, internet access, microscopes, hand lenses, computer models, GPS, field journals
- Setting up and taking down equipment
- Peer collaboration
- Organizing
- Finding patterns
- Making inferences
- Predicting trends
- Collecting and recording data using the appropriate units
- Expressing data in tables and graphs
- Constructing and labeling drawings
- Using formulas to make predictions
- Forming a hypothesis
- Developing an experiment
- Following directions
- Teamwork and communication between people
- Reading graphs and tables

- Writing coherently so others can read what you are communicating
- Deciding what is essential and what is not necessary to solve problems
- Determining independent and dependent variables
- Making calculations
- Cross-multiplying
- Peer tutoring
- Using technology like probe-ware and simulations
- Using the internet to research information
- Demonstrating safe practices during laboratory and field experiments
- Being able to respond to accidents where injuries, broken equipment, or fire occur
- Understanding the use and conservation of resources
- Demonstrating proper disposal and/or recycling of materials
- Understanding the limitations of science
- Being able to make and build a hypothesis
- Being able to distinguish a hypothesis from a theory
- Being able to show the difference between a theory and a law and that theories are used to explain laws
- Learning how to use safety equipment properly
- Learning patience
- Learning how to measure mass, length, volume in both the metric and English system with precision and accuracy
- Being able to convert from one unit to others
- Seeing how scientists and engineers learn things for the first time
- Proving equations or laws

- Processing information to see if it is relevant or not
- Using logical reasoning, experimental testing, and observational testing to critique scientific explanations

These are just a handful of skills students can learn while conducting investigations. Imagine the skills they would never get a chance to develop if not allowed to do investigations. We do a great disservice to students when we hold back investigations and rob students of opportunities that would make them more marketable and competitive for spots at colleges and in future careers.

Why would you not want students to learn these investigative skills while they are learning your content? Many times, teachers say they do not have the time to teach all the concepts and skills in the allotted time they are given. I hear teachers say, "I don't have time to do the labs." I say, "I don't have time to do a lecture." **Lecturing is a very inefficient use of time.** Do you really think those students are fully engaged in learning while you are talking? You can be half asleep and take notes from a teacher's lecture (I know I did). Just look and see how many times you have to wake students up during a lecture. Do you really think that most of your students are going to study those notes for the test? If they did study, they would only pass the test if written at a low rote memorization level anyway. If they do not study it, they will remember next to nothing and have learned no observation, problem-solving, or critical thinking skills (nor anything else listed on pages 23-25). If you have students learn their content through investigations, they will have acquired and practiced science processing skills, not to mention they had physical experiences of the content and concepts that make it easier to remember. They don't have to study to show their skills and remember their experiences. Why would you purposely deliver content to students in a way that will not prepare them to be more successful? Having students learn through investigations, in many ways, makes the class easier for them. You cannot control what students do outside of class, but you can control what happens in your lesson. I will not lie to you; what I will tell you to do will require a lot of work upfront. But it will be less work on the back end. The workload will get better as time goes on.

Later, instead of constructing your lessons from scratch, you will only modify them. If you work on a team, this will be a piece of cake because the preparation can be shared with teammates. You need the guts to try teaching through investigations. Let go of what has not been working and what has been holding your students back. The students who are serious about learning will be thanking you. I know this because it is so rewarding to hear how your class is so much more exciting and engaging than the other science teachers who did not put as much effort into theirs. It does take some convincing with some students at first because some may not be used to thinking and learning this way. At the beginning of the year, I tell my students that my goal is to reprogram their brains, and in the process, they will learn some physics, chemistry, or biology. There may be struggles at first for some, but later they will get the hang of it as they get more familiar with the process.

One of the biggest reasons why we should teach science this way is because this style is best for Gifted and Talented students, Regular Education students, and Special Education students at the same time. Differentiation is built-in at so many places just by the way the lessons are organized. You could potentially have all three types of students in the same class doing the same activities, even answer many of the same questions, and have it all still be differentiated. The students will be in groups or pairs to decide what role(s) they will play during each activity. Often, roles will change with the activity, and sometimes they will not, but it will be about their choice. Many of the questions asked in the investigations are open-ended, so students can answer them with as much or little detail they want or is possible for them. Projects also allow students to work to their ability even when the directions are the same for each student. Look at how the students learn when they investigate phenomena, information, and ideas. The learning is full of places for differentiation to be used that are naturally available. Students have opportunities to describe all of what they predict, observe, and understand. They do this at their level and with as much detail as they want to. Students can take on different roles in investigations through gathering data, showing each other how to collect and analyze that data so it is more precise and accurate, and understand why the information is saying what it is.

Gifted and Talented students will have the freedom to go into as much detail on their explanations in their labs and projects as they want; just as the Special Education student will explain their understanding of a concept or work to their ability without the activity being overly demanding on them or limiting them from achieving more. With all the differentiation that is inherent to this style of teaching. There is still room for teachers to add more where they see the need. Having used this style for years in a mixed ability classroom, I will tell you that there are times where that

Special Education child will outperform those Regular Education kids. I have seen it many times because many Special Education students' weaknesses do not hinder their learning. This style is so good for students who have reading disabilities. Students get to practice their literacy skills while at the same time not having to be reliant totally on them to understand what is going on. They are defining their terms not with written definitions they would have to read but with actions that they can observe to build their meanings in the context of the other content. Why would you not want to teach this way when so many students from all walks of life can benefit from learning through investigations?

The last reason why we should teach through investigations is that our state standards tell us to do it. In Texas, like most other states, the very first thing that is mentioned in each science subject's standards is "Students conduct laboratory and field investigations, using scientific methods during the investigations, and make informed decisions using critical thinking and scientific problem-solving." The Texas Essential Knowledge and Skills (TEKS) go to great lengths to define what science and science inquiry is; and how it is done. The wording implies that students should be learning their science concepts through scientific inquiry. Why? Because this is what research shows works. They even say, "Scientific methods of investigation are experimental, descriptive, or comparative. The method chosen should be appropriate to the question being asked" (Texas Education Agency, 2009). This phrase says that the appropriate way to have students learn scientific skills and concepts is by performing an experimental investigation, a descriptive investigation, or a comparative investigation. The TEKS did not say to give a lecture through PowerPoint. If lecturing through PowerPoint is just as good as investigations, they would have mentioned it here. But they did not, and that speaks volumes to me. The state law that our lawmakers passed tells us what teachers are to teach and tells us how teachers are to teach students in science classes.

What I am talking about doing is not new. There has been a movement for a long time to do this. The College Board has changed the way they want A.P. Biology taught and tested. SAT and ACT are also making similar changes. The National Research Council released *A Framework for K-12 Science Education* that agrees with what I am proposing. Many districts are using this document to shift how they are delivering science instruction. These changes are due to needing to let go of things we think we need to do like lectures, PowerPoint, or notes; and really rely on investigations. I have been able to teach this way with great success and have streamlined and simplified what many others make more complicated than needs to be.

Chapter 3

What Are Investigations?

There are four categories of investigations: **descriptive, correlative, comparative,** and **experimental**. Modeling and pattern-seeking are types of investigations used in any one of the three categories of investigations. Fieldwork, secondary research, and identification and classification are included in **descriptive investigations**. The controlled experiment or fair testing is classified as an **experimental Investigation.**

A **descriptive investigation** is where people investigate questions through observations of phenomena recorded to quantify or qualify something without manipulating any variables. Fieldwork is a type of descriptive investigation where students study nature through observation. Examples would be when you study animal behavior or a black hole. Pattern-seeking is involved with fieldwork by recording natural events or carrying out experiments while looking for patterns in data where the variables are not easily controlled. It is important to note and record variables to identify patterns that may occur between them. Pattern-seeking can happen in almost any investigation like modeling. The ultimate pattern-seeking investigation is the study of the weather because so many variables are being observed simultaneously. Secondary research can also be classified as a descriptive investigation where students look at data or information from previous research from someone else. Secondary research involves gathering and analyzing other people's opinions or scientific findings to explain observable events. An example is anytime you look up other people's work and try to learn something from it. The last type of descriptive investigation is <u>identifying and classifying</u>. This type of investigation involves sorting objects, events, or concepts into groups or categories. Classification Keys can be created or used to do this. Trying to classify or identify different insects or leaves are examples of common investigations of this type.

Correlative Investigations find positive, negative, or no correlation between variables. No one manipulates any variables; they just look at data to see if there is any correlation between two variables. Investigators can look at natural observations, surveys, documentary research, and archival data to see correlations.

A **comparative investigation** is where students compare and contrast objects or phenomena (happenings). Compare two or more groups on one variable to determine the strength of a correlation between the objects or phenomena being compared and the variable tested. This type of investigation involves collecting data on different items under different conditions. It is like experimental investigations but does not have a control. Seeing which plant can absorb the most carbon dioxide to help control climate change is an example of an investigation of this type. It goes through the experimental process of observation, question, hypothesis, then testing the variables through an experiment (but there is no control group), then data is collected, and analysis from which we make our conclusions. Since there is no control group, you also do not need constants. You are just comparing how the independent variable reacts to different situations.

Experimental investigations use the scientific method where students use experimental and control groups to test a hypothesis. This type of investigation tests the effects of an independent variable on a dependent variable. Experimental investigations use the usual science process. The sequence is first a question, then the hypothesis, then the experimentation with the experimental and control groups, then the data collection, analysis (pattern-seeking), and finally a conclusion. It uses control(s) to test the relationship between the two variables during the experiment. All factors are held constant except for the independent variable. This type of investigation is also known as a controlled experiment or fair testing. Here we manipulate the nature being studied while testing the causes and effects of variables to understand the science content. We find vaccines for viruses using a controlled experiment. A drug (the independent variable in the experimental group) is tested to see if people get immunity. A placebo that does not contain the drug is given as a control to compare to the experimental group. Everything else between the two groups must remain the same; only the independent variable is different between the experimental and control groups. Everything else remains constant.

Modeling can help people understand how processes work or explain ideas or concepts not easily observed in nature. It is a simplified representation of a system. Models are useful for studying things that are too big, too small, or too dangerous to explore directly. A simulation is where you imitate some real thing, situation, or process. A model can be a simulation and used in any of the three types of investigations. Students can construct or use models. Examples include posters, pictures, sculptures, animations, and computer simulations like PhET or ExploreLearning. Computer simulations are an excellent way to have students perform investigations without getting out and setting up lab materials. You can also conduct experiments that may be too time-consuming or costly by using simulations. Simulations should never replace a hands-on lab but can be used as supplemental activities to support it.

The easiest thing for teachers to do is find investigations that have already been constructed that fit the content and the context of the content you need to teach. Seven Sides Publishing has already done this for you with Simple Investigations for Biology, Chemistry, and Physics. The best thing to do for your students sometimes is to build or modify investigations so that you can tailor them to fit your students' needs according to your state and district curriculum (but this is much more time-consuming). We have digital versions of the labs available to modify them for your wants and needs. If there is content that needs to be covered where no investigations seem available, you are in a time crunch and do not have a team to work with; you can have the students do research using broad or open-ended questions to find the information you need your students to know. The more effort you put into building these investigations, the more your students will get out of them and enjoy doing them. Keep in mind what skills you need your students to practice while acquiring the content (found on pages 23-25). It does take time to accumulate the number and types of investigations you will need to teach your class on your own. So Seven Sides has helped you out with lab manuals and guides to virtual investigations you can use.

Chapter 4

How to Teach Through Investigations

If you watched the movie, the Matrix, you might remember the part where Neo, the protagonist, first met Morpheus, his teacher. Morpheus told Neo, "No one can tell you what the Matrix is; you must see it for yourself." Just like when Neo was trying to learn about the Matrix, students cannot be told what science is; students have to experience it for themselves. They have to go through the processes and be immersed in them to learn how to manipulate them as Neo did. Neo went into the simulations to practice the situations he would later face. Neo going into the simulations is like the students doing their investigations both live and in computer simulations. Neo would then be tested by the Smiths, who were programs that tried to control or kill him. When he faced them, he would learn even more, like your students, when they take tests that require higher levels of thinking or the more important tests of life. When students are continuously challenged, they will learn and adapt to the level at which they are working; and quickly exceed expectations. It just takes time and practice. Neo once asked Morpheus, "Are you telling me when I am ready; I can dodge bullets?" And Morpheus said, "No. When you're ready, you won't have to." He would be able to manipulate the whole Matrix just like the students will manipulate and use science. If you saw the end of the movie and the rest of the trilogy, you got to see the unbelievable skills Neo was able to exhibit. Our students are not that much different. No, they will not be able to break the laws of physics but think of what our students may be capable of in their futures if they learn the content and skills that would make them great thinkers/problem-solvers needed for the 21st century. Think how much more marketable they will be if they acquire skills that would allow them to think on their feet to solve new complicated problems.

What I will explain in this chapter is not hard to do and has lots of flexibility. Avoid all lecture and low-level thinking like memorizing and rote activities, especially vocabulary worksheets, which is nothing more than playing science trivial pursuit or reciting the textbook. Teaching through investigations does not mean you will not talk to your students; you will have different conversations to act more like a facilitator/guide. You decide what investigations you want to do and how involved you want them to be. The only thing that will limit you is your will, imagination, and time to find and develop student activities. Seven Sides Publishing has prepared lab manuals with investigations in them. They are called Simple Investigations because the investigations are easy to read, prepare, and perform. I know your time is valuable and in short supply, so they were written in a way meant to be quick and easy to implement. There are separate lab manuals for Biology, Chemistry, and Physics. They contain a guide to show you where you can find related virtual investigations that complement each section's labs. Together there are more than enough resources to have students perform investigations every day. There are also Unit Concept Maps students can use to help make sense and organize vocabulary. These all have editable digital versions available if you would like the freedom to modify and personalize them.

What follows is an overview of how to have students learn science through investigations:

1. **Look at state standards and local curriculum** to see what concepts you need to teach and in which order. You can find these on your district's and state's educational websites. I like to keep a hard copy nearby, so I can refer to it when I develop a curriculum for a class I am teaching for the first time or adapting my curriculum to newly changed state standards. The Simple Investigations lab manuals follow a learner-friendly scope and sequence. You can follow that if your district does not have a mandated scope and sequence.

2. **Build concept maps.** Some examples are shown on pages 42-44, and others are provided for you at the beginning of each section in the Simple Investigations lab manuals. Digital versions are available to customize to your State, district, and school. These need to show how all vocabulary from the local curriculum and state standards relate to each other using icons and visualizations. Arrange the vocabulary to show how it can be organized to bring forth the meaning of the content to the students and give them an overall theme for the unit. Some units are easier than others. I like having them on legal paper, so students have more places to write notes in the white space. If content gets too much to put in them, break it up into two pages using the front and back. You do not want them to be too busy where students don't have room to write notes on them. Put things on there that

students would typically have to memorize, like formulas and constants that they can use to solve more complex higher-level questions. I let my lower level-classes use these while they are taking their tests. My more advanced students should use them during their review and put them away during the test. Concept maps are not a necessity to use when teaching through investigations. But they do help students organize their thoughts to think more like a scientist. There are many ways in which you can use them. I like to hand them out at the beginning of each unit and tell the students they can write any notes on them they want, including definitions. At the end of each day, you can have students summarize what they have learned on parts of the concept map. Before the review, I like to have the students finish filling them out at the end of each unit, writing words on the arrows showing how the content relates to each other. I like to say when all the information is there, the concept maps should talk to the students, but students must develop the skills to read them. If they are not used to them, they feel uncomfortable the first time they use them. The more practice they get with them, the more they like them and rely on them. When they use them during a test, they allow students to learn as they take the test. Using the concept maps, they can explore new applications and make new connections they may not have during the unit. I am sure there may be more ways you can find to use them. I found them helpful with this style of learning.

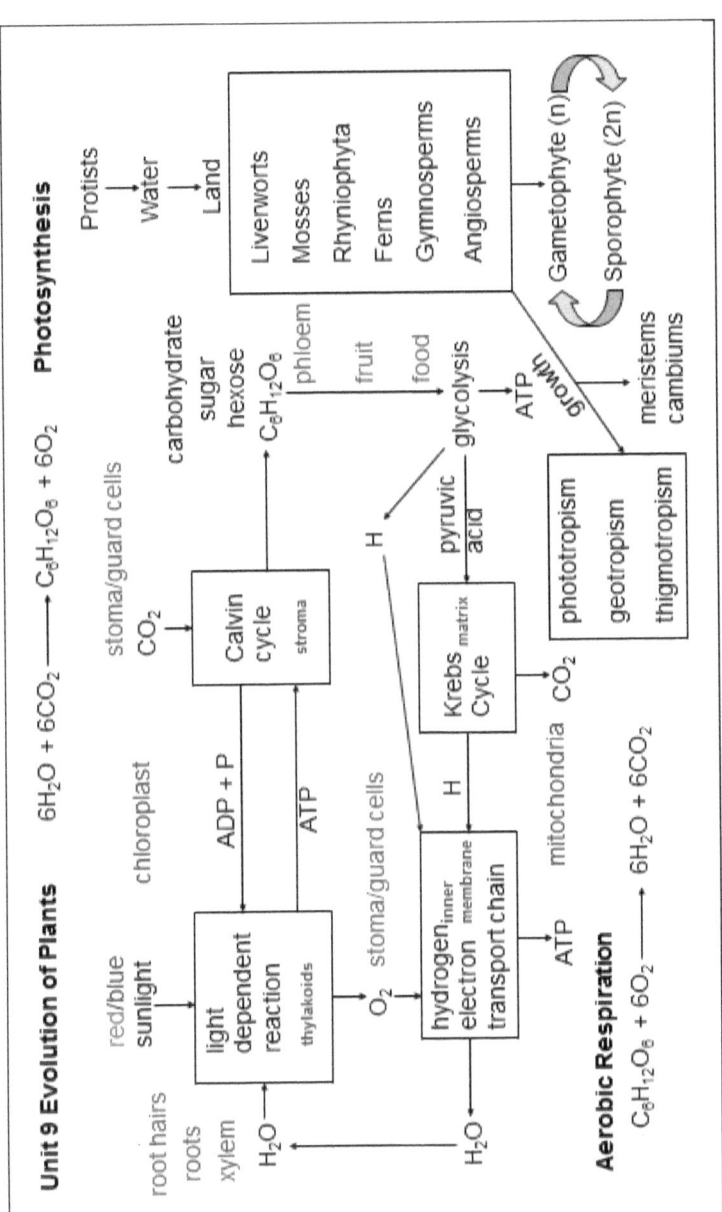

Unit 10 Circuits

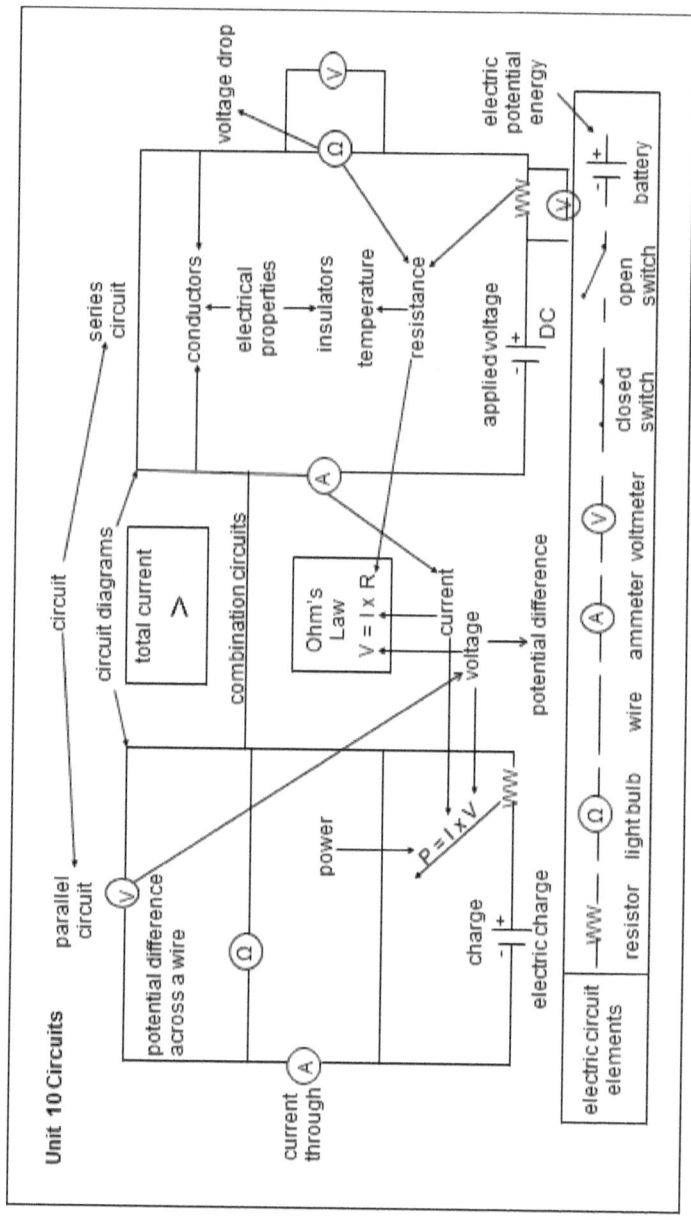

Unit 9: Energy in Chemical Changes

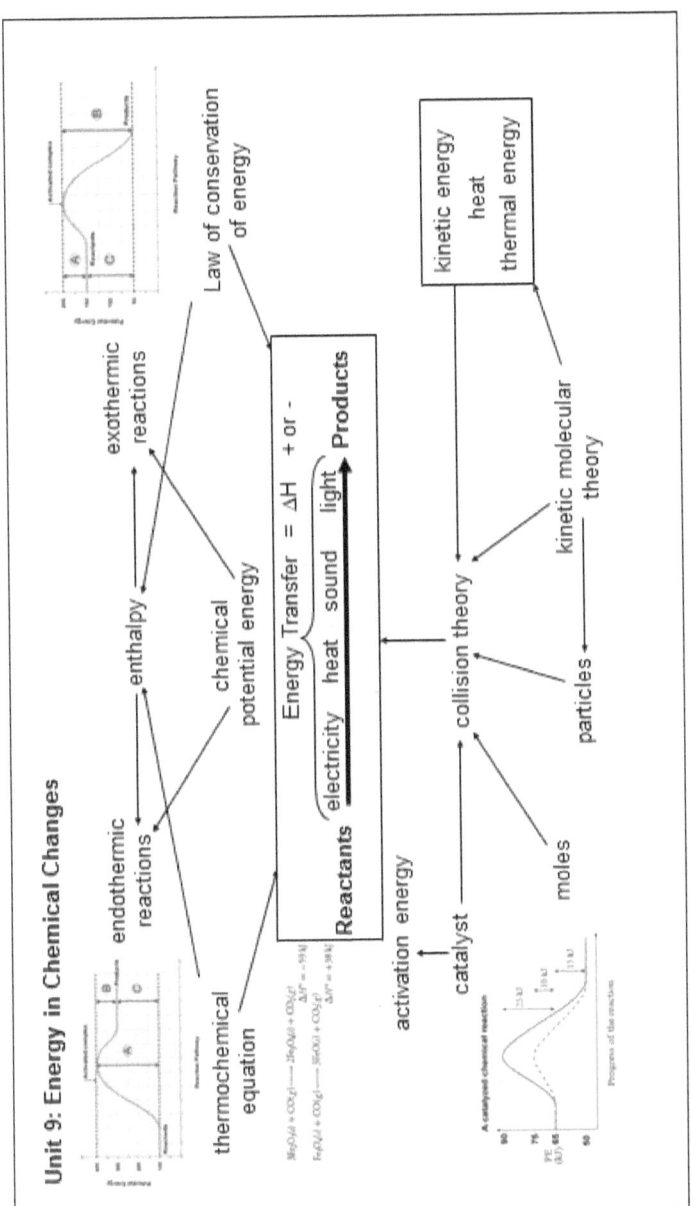

3. **Build tests for each unit** that do not have rote memorization questions but questions making the students think and use the concepts and science process skills like those on the End of Course Exams, standardized tests, and college entrance exams. You can search the internet for released test items for different state and national tests. The States have their own released test questions you can use. I found a pretty cool website with many of the states' End of Course Exams collected together. It is called EDinformatics; the current URL address is http://www.edinformatics.com/testing/testing2.htm . There may also be sites your district pays for as a one-stop-shop to pick up all kinds of test items that include SAT and ACT prep. The tests you make do not need to be large if the questions are done well. The majority of my tests are 30 questions or less. Use them as they are, or modify them to fit your needs to use vocabulary you need to be assessed. Your district may have already organized a bank of test questions for each subject and unit to make this process easier for you. You can build these types of questions too, but they are very time-consuming to make but could be well worth the while. You may want to do this with a collection of people to make sure your questions ask what you want them to, at the appropriate level for your students.

4. **Find labs, virtual labs, investigations, and activities** that make the students think, observe, and research the content in a way that mimics the style of the test you constructed. You want to find investigations that

introduce the content you want to be taught and yet at the same time practice different science processing skills. Refer back to pages 23-25 to get ideas on which science processing skills you want to be practiced. Seven Sides Publishing provides plenty of resources through the Simple Investigations for Biology, Chemistry, and Physics to help you achieve this. They are easy to read, prepare for, and implement. There is a wealth of hands-on labs and guides that show you where to find virtual investigations.

5. **Modify the investigations and activities** where needed to fit your students' objectives, content, and tests. Digital copies of the Simple Investigations are available with a minimum purchase of lab manuals or a digital license. These digital copies will allow you to edit any of the labs to fit with your equipment, your curriculum, or allow you to make any other modifications. Many of the virtual investigations we direct you to can be edited; others are used as-is.

6. **Build new activities and labs** to have students investigate content you could not find that matched the content you need them to learn. I tend to look at which standards I did not cover and use the wording of those standards to build an investigation. If you use the Seven Sides Simple Investigations, you should not have to do too much of this. Feel free to do as much building that you feel inspired to do. The more effort you put in, the more your students will benefit.

7. **Locate and acquire the equipment** needed to complete these investigations. You may have a great lab activity, but the students cannot do it if you do not have enough equipment. I have had to do demos for a while until I could acquire enough equipment for a class set so the students could do the investigations on their own. Students are usually more engaged in the investigations if they perform them instead of having teacher lead demonstrations. Sometimes when I have limited equipment for different labs, I have the students rotate lab stations to do their labs at different times. They all eventually get a chance to use each piece of equipment and do all the labs, even if I don't have a class set for those labs. Make sure the wording is clear to the students in those labs, or you will be spending a lot of time scurrying around answering questions to clarify information at the different stations.

8. If you need to practice math applications, **find math practice activities** that will follow the investigations that prove the math objectives required by your state and district. If there is any math practice in your class, have them do this near the end of the section. Your investigations should show what the mathematical relationships are between the variables during those investigations and experiments. After you prove the formulas or show why they exist through investigations, have students practice solving the simple problems first, working on the more complex problems last using all of the formulas that your curriculum requires. At the end of

the unit, use more complex problems that integrate all of the formulas or concepts. Students also need to know how to solve complicated word problems. The teacher will need to model and have the students practice finding what information the problem gives them; and what the question is asking. Have the students make a list of all of the variables given in the situation. Then have them find the variable being asked. Many times the information provided is not mentioned explicitly in the problem. For example, when a ball is being dropped; or a car starts from rest, the initial velocity is zero. Students need these modeled for them to learn to see those hidden variables that were given to them. Once they know the variables involved, then you can show them how to find the right formula to use. Typically, the formula they will start with will have all of the variables but one given in the equation. If it is missing two or more, this formula cannot be used yet. Many problems require multiple formulas, which means they need to know which formula(s) can give them the variable for which the question is asking. Many times, there is more than one way to solve complex problems. Solving the same problem in multiple ways could be used to the student's advantage. Students could solve a problem in various ways to see if the answers correlate, making sure they did not make a silly mistake somewhere; this requires lots of modeling and practice. I have found this to be one of the most challenging skills to get students to learn. Many of the investigations Seven Sides Publishing guides you to; shows you how to find or prove the formulas in the laws students will use. (TIPERs) Tasks Inspired by Physics Education Research (sense-making tasks or ranking tasks) are a great place to get some good practice, make students think at a high level, and reorganize content in the context of other content. They are well developed for physics, and some in chemistry are worth using. After seeing how they are structured, I know they can also be made applicable to biology. However, only physics and chemistry sense-making tasks or ranking tasks are currently available ready to use.

9. **Modify or create math practices** you could not find that your students need. Mimic the types of questions on your tests.

10. **Build reviews/study guides** students can use to practice the concepts and skills that mimic the way students will need to think to be successful on the test. They should not look like the tests or be the test (many teachers do this); they should make the students think the same way, going through the same processes and skills, but be different from the test. Have the students work together while going through them; this will allow collaboration and peer tutoring opportunities. Watch out for students that are trying to get answers and avoiding the thinking they should be practicing. The processes students go through when solving problems are way more important than just having the correct answers. Students not worried about the process and just

trying to get answers will not do well on the tests. They are accomplishing nothing by just filling in answers if they do not practice the process of thinking and problem-solving.

11. **Make sure you (the teacher) do all investigations, activities, concept maps, and reviews as the students would do them** to know what the results should be, build the answer keys, and anticipate problems the students will have; this will help you make changes where needed before the students attempt the activities, so their learning experience is as smooth as possible. Sometimes when you have the students perform activities, those who do not get what to do may sit there doing nothing. Or they are performing the wrong actions. When you do all the activities, you already know what they should be doing and recognize when they are doing anything wrong.

12. **Set up your classroom in a way that promotes scientific collaboration.** Have them sit at tables in pairs or groups of three or four, considering students' abilities and personalities. Make sure students have room so they can safely move to perform experiments, and you can get anywhere in the room in just a few steps to keep close proximity to your students. Make sure you create a seating chart the first day and do not let the students seat themselves. If I notice whole groups or partners not performing well, I try to move students until I see success throughout the room. If you allow your

students to seat themselves, discipline problems will quickly show up most of the time because challenging students seem to be naturally attracted to each other like magnets. I like arranging my students in pairs, which can also be grouped with another couple in groups of 4; this is so students can collaborate. Having students discuss what they are learning imprints the content in their brain the more they talk about it. If students verbally talk about what they experience, they are more likely to retain that information and access it easier in the future. If you promote language-rich environments that include the specific vocabulary in the investigations, this will increase student retention and demonstrate understanding on assessments. This process also builds the "soft skills" to articulate their knowledge of content and skills. Every job/career on the planet requires these skills. You want to provide opportunities for students to teach each other with some peer tutoring. As long as they have on-task discussions, you will be amazed at the learning that can take place. So arrange your room so that positive collaboration can happen while limiting extra social talking with other groups.

13. To **teach the content,** I usually start by passing out the concept maps, so students can see right away all of the content they are expected to learn. I use those to talk about what activities they will be doing to experience those concepts for that unit. Next, I have the students start their investigations that I have planned, followed by any math practice that

needs to take place to practice problems that go with that specific content. Then I have students move on to new investigations that cover new content in that unit, followed by the math practice (if needed). Make sure the rigor of the content increases as the unit moves on from beginning to end. Start simple, then work your way to having students apply the information to solve more complex problems, integrating all of the content for that unit; this is super important, especially for struggling students, to experience success early. This progression keeps them engaged and builds their confidence so that they know they can do the work. It is crucial to instill a growth mindset from the very beginning. A growth mindset is where a person believes that their brain changes as they learn. They may not be good at a particular skill. But with practice, they can make their weakness into a strength. This practice allows the students to know they will be able to do new things they could not do before. They may struggle at first. But that struggling is not only typical but many times beneficial for learning. Most of the time, all you have to do is start the process by walking them through it individually by modeling it for them. **Stay on your feet** to find those students who may need guidance by looking for students doing nothing, looking confused or frustrated, or performing actions you know should not take place. If you see any of this happening in multiple students, you may want to reword or reorganize that part of the activity, confusing them, so it will not confuse your students in the future.

When I have finished all of my unit's activities, I like to go over the concept map. You can have the students connect the concepts on their concept map by writing notes to speak to them as they go over each activity. Working on their concept maps may be an excellent way to end class each day by summarizing what they have discovered on their concept map that day. Ideally, I like the students to tell me what they think, and I make corrections to their understanding of the content as needed. When summarizing the concept maps before the students do their review, the class looks like a lecture is going on, but it must be interactive. Lately, I have used google classroom to have students fill in concept maps together as I explain them to the students. Working together forces students to participate to have the concept maps talk to them. The students must participate in this process, or else you will lose student engagement. After they have their concept maps completed, they work on their reviews in groups; while referring to their concept maps. Because you can control what goes on in your classroom but not what goes on at home, try to leave time for them to complete this in class. Those students that do not finish in class need to complete it at home. Students, who sit there doing nothing, saying they will complete it at home, will not. So, encourage/make them do it there, in class, where you can still help them. You cannot help them when they are at home when they have questions. Also, students should look over the investigations they did for that unit to review the content. On the day of the test, let the students ask questions about any content they may not understand from the review. If no one asks questions, ask them questions over the content they seem to have difficulty with to stimulate their thinking. After that, give students the test, and if they all finish early, or if you have time the next day, have some enrichment. The enrichment must morph because student's trends and interests keep changing. It could be a fun lab/activity or a video that goes into the content at a deeper level or tells a story about real-life applications of the content that they just studied or are about to learn for the next unit. I have used Through the Wormhole, Cosmos, Planet Earth, Trials of Life, Brain Games, and Superhuman Series that have interesting stories and content. Students can see how this content is applied to real-life with more significant ideas.

You must implement the activities while you are **on your feet**, constantly monitoring students' learning. Try keeping close proximity to all of your students, so they don't think they can take a break while you are with others. Make sure you have a fast path to anywhere in the room, no matter where you are. Make sure that when you are talking, you are only giving directions, asking or answering questions, summarizing (but this must be very short), or building relationships. Anything else will not help them to think. Try to learn and become good at Socratic questioning. Merriam-Webster says Socratic questioning generally involves the use of Socratic induction, a way of gradually arriving at generalizations through a process of questions and answers, and Socratic irony, in which the teacher pretends ignorance while questioning his/her students skillfully to make them aware of their errors in understanding. Socratic questioning keeps the students from depending on you for all the answers and models how they should think about topics and problems.

14. When you have tested the students, **analyze the results of those tests**. Go over those results with your students; so they know what concepts they did well and which they did poorly, and reteach simple mistakes that can be fixed with different strategies.
15. **Reflect on how well the unit went**. Go back and modify investigations, practices, reviews, or sequence of delivery of your developed curriculum. Spiral content that students have a hard time conceptualizing back into later units where appropriate. You should also always evaluate how well the students took in the content and make notes or adjustments right away (right after you have done the activity) while it is fresh on your mind, so you get it changed for next year. Otherwise, you will forget, and your students will keep making the same mistakes over and over. Reflection is a quality of master teachers, and this process is beneficial for improving instruction for future students. It is also good to do with students to modify behavior, instill confidence, and show proof that they are progressing. I like taking time the day after a test to

reflect with my students. Be forewarned, students who do not care about their learning will hate this time, so this must be done strategically, choosing your words well. Many students think that once the test is done, they will never see that content again. They have experienced this in the past, and it is not true for most science concepts in most science classes.

16. Somehow **document and save how you have implemented your curriculum,** so it does not need to be developed again next year. The more years you implement the curriculum, the less time you will spend building and modifying it each year.

- Notice there were no vocabulary definitions, lecture notes, or memorization of any kind. Students should be taught how to use the content, not how to memorize it. Many call the strategy we want to implement **"Learning by Doing."**

When teaching your content, introduce it through investigations. Like real scientists do the first time they discover or prove something. You want the students to think like a scientist as much as possible. They not only develop their critical thinking skills but also learn how the process of science works. Knowing how science works will help them understand and appreciate the process when discoveries are made and have confidence that science is not about just making things up but about observing the world around them. Going through the science allows students to build their content knowledge in the context of the other content as they build relationships between pieces of information;

this also engages students in a way that allows you to see how they are involved in their thinking. You can see that the students are engaged through the actions they are performing and their conversations. You want them always working with a purpose to an end.

Investigations almost always start with some type of question that gets the brain thinking. By looking at the student's actions, hearing their conversations, and looking at what they produce, you can observe what students understand. Knowing what they understand is essential to see, so you know what you need to communicate to the students next. This interaction will also be important to your administration and evaluators when they walk through; because they will also observe the learning. Seeing your students engaged (and they will be engaged) will be commonplace if it is not already. Students not involved in learning will stick out like a sore thumb to you and the other students because they will be the ones doing absolutely nothing. You should see student physical actions that are observable that you cannot see during a lecture.

Public education aims to build students who are effective communicators, competent problem-solvers, self-directed learners, responsible citizens, and quality producers (the outline of my last district's Portrait of a Graduate) through student engagement. All educators, especially administrators and teacher evaluators, love engaged classrooms, which you will have every day just because of how students spend their time in this instructional model. Research has been done where the vast majority of secondary students describe that school as "boring." What students are doing should be far from boring, especially if the investigations you find and build are implemented to the best of your talents and abilities. Having students learn content and skills through explorations requires your students to be engaged; if they are not, nothing will be happening, and that should be easy to spot. This instructional practice keeps students engaged and results in students being able to competently problem-solve, in a way, that our country needs to fill jobs and solve problems for the 21st Century.

Each year teachers struggle to get to all of their content because many state standards are so big, they do not allow time for critical thinking. We might as well have our students engaged in activities that maximize learning and gets students thinking on their own at a high level from the beginning. Many state standards and curriculum are a mile wide and an inch deep. The process of teaching through investigations allows you and the students to cover a lot of content while still ensuring skill practice where students must think critically and get into more profound understandings. Multiple places in each unit have authentic assessment opportunities to demonstrate their knowledge, skills, and learning in many ways. The Simple Investigations lab manuals have lots of resources to help guide you through this process.

Chapter 5

Solving Systemic Challenges You May Face

Lack of Equipment

Many of the schools where I taught did not have the equipment to do all the labs I needed. I had to come up with creative ways to get the materials I needed. I like to find materials that are safe but cheap. Buying toys at dollar stores and Wal-Mart help keep the budget from getting too high. I have also had a lot of luck going to Good Will to get the stuff I needed. You may be able to go to companies that may be interested in sponsoring you or giving you equipment they are not using anymore. I have received equipment from labs that closed or received new equipment and were willing to donate their old equipment. Just make sure that you inspect the equipment you receive from them; you do not want something to short out while the students are working with it. You may even be able to get in touch with a church that may want to sponsor a wish list. One year I applied for a grant, which gave me a lot of fast resources. Those monies are out there. But watch out for who has access to the funds. The grants usually stipulate how they have to be spent. Just make sure that what you spend the money on is specified in the grant.

For Chemistry, I try to use household items to do experiments; this limits the dangers you will face in the classroom and costs much less. Catalogs are overpriced. Walmart, dollar stores, and Good Will are not. Most labs that call for hydrochloric acid or sulfuric acid (not all labs) can be replaced with vinegar. Vinegar is not very dangerous and is easy to get, has a good shelf life, and is easy to store and discard. Using more household chemicals may also help keep your Material Safety Data Sheets (MSDS list) more comfortable to manage with all the rules and regulations we must abide by to keep our schools safe.

Safety

At the beginning of the school year, there are always lab safety activities before the students start learning the class content. Middle school students need to have more lab safety lessons because they have not had much time in labs. The higher-level high school science classes need less time to spend on lab safety at the beginning of the year. Before each lab, prepare your students for the dangers and precautions they will need to take. Each Simple Investigation starts with the setup information, including the materials students will be using. At the end of this first section, there is a question to evaluate the lab's safety precautions. It states, "Looking at the materials and lab we will be using, what are the safety precautions we should take to protect ourselves and materials during the investigation." This question is where the students and teacher need to look over the lab and talk about all the lab safety students need to be aware of when doing that lab. Do the lab safety teaching when you start the lab, specifying each lab's dangers. Assess all your students at the beginning of the semester/year with a safety quiz to see who knows lab safety, who struggles, and how they struggle. Explain those items the students typically struggle with to the class. Let your students know there is absolutely no eating or drinking in the science lab. There are too many dangers, and you do not want students consuming things that could make them sick. It is particularly dangerous when you look at all of the food allergies that are out there. Students can have an allergic reaction from the residue left by food another student ate in a previous class. So no eating should be taking place in any class, not just science classes. Show where the safety and emergency equipment is located in the room, along with any school and district emergency protocol. Have them and their parent(s) sign a safety contract to know they will be held accountable for their behavior because of the dangers in the science lab. Do not take safety lightly; be vigilant and stress safety within the labs. Safety training should never stop and is enforced every day so that students don't start thinking there are different rules for different days. If you make an exception one day, they will bet they can make another exception again, and all other rules are now subject to debate and exception.

Grading

Grading is the biggest problem I found in teaching through investigations; you end up with lots to grade. Pay attention to the grading tips I give here if you don't have very many grading tricks. Much of the grading can be done in class as you monitor student learning, or you can choose not to grade certain things at all. Just don't tell your students that you're not grading it until they have already finished the activity. I do not know about you, but grading papers is what I hate doing more than anything else I have to do as a teacher. Even so, I usually have student's papers given back to them and grades in the grade book by the next day. I believe you need to try to have all of your student's work graded each day and give students their graded papers back by the next day, so you do not let grading pile up on you, and students get feedback as soon as possible. You may even want to develop a system where you require each student to let you look at their work and show them mistakes before they turn in their work, so you can give them immediate feedback, show them what they are doing wrong, and let them correct their errors right then and there while it is fresh on their minds. When I see that they have everything correct, I put a red check on that side of their paper, telling me that I have already checked it and that everything is right. When they turn it in, you have already graded it and don't have to grade it at home. Some students want to get it done and do not wish to have you look at it (and one or two slip by), so they turn it in without you looking at it. Because most of the class will do what you ask, most of the papers are already graded, so there will be very little to grade at home. Be careful about absent students copying other student's completed work that you have already turned back. If this starts happening, and it will happen, you may want to hold the graded papers back until the make-up work is turned in for that assignment. Some teachers may not want to turn any work back because of the amount of cheating that happens. You would have to take into consideration late work policies at your school to make the best decision.

You can also incorporate what I call the **Captain System**; this is where the students work in their groups and race to see who first finishes their work correctly. You grade the captain's work, and they check the others in their group. They cannot let the other students see their paper(s) or tell them the right answers. But they can say to them what they have wrong and have the students try again. What is neat is that each student usually gets to be the captain at some time, if done regularly. If someone is struggling, students can provide tutoring to help them out without giving them the answers. Stress with your students the right answer should not be the priority, but the process they go through to get the right solution and know why it is correct. Hopefully, the right answer tells you that the student went through the proper process. When the students are all done, you only have to correct a fraction of the class's papers, and everyone should have gotten immediate feedback, should have received a 100 in the grade book, and you should have nothing to grade at home. Just watch out for cheating or friends giving other friends answers, or captains not doing their job. When you get all students to buy in, it saves you lots of time grading, and you get some peer tutoring out of it. It benefits all involved, not to mention the captains are evaluating, which is very high on Bloom's Taxonomy.

If you want to go a step further and mix it up, here is another version of this strategy. If you want to get more students active (you know who they are) to use some positive energy, you can give them the "green light" to move around the room and correct other students' work after they have correctly completed theirs and you have checked it. You can pick as many students as you want, from one student to the whole class. You could even have any student who has all of the right answers check someone who still has not finished getting their paper checked until all students have the activity finished correctly.

There is a new trend where some educators do not grade anything while students are going through the learning process. They only grade assessments. I noticed at-risk kids refuse to do work if they know they are not being graded on an activity. They have a hard time seeing how putting work into something that is not graded will cause their grades to go up somewhere else. Some do not know the connection between effort and good grades on a test. Watch your students, and know them to see how much you really need to grade. There is a feature on Gizmos from ExploreLearning, which gives students a five-question quiz after they do the Student Explorer. The Student Explorers guide the students through a series of investigations that introduces science concepts to students. If students skip lessons, they will do poorly on the assessment because questions can be quite complicated. When they work hard, they tend to do better; this gives feedback on how effort and hard work really makes a difference. The website grades the quizzes and sends them to me immediately. Google classroom has features you can take advantage of where you can set up lessons there, and it can grade everything for you. When you find a system where suitable quality lessons are given and the site grades stuff for you, you have a bonus to help you save time.

Need Time to Organize

This process will be time-consuming to set up, so you may want to start organizing during summer vacation and take time to modify and organize while on Thanksgiving Break, Christmas Break, or Spring Break. These breaks are when I get lots of work done when starting new classes or making a significant change. It will take a lot of time upfront, but this gets much easier once you have gone through the process. After that year, all you have to do is fill in gaps and make modifications in things you did not like, or change to new stuff that looks cool that you come up with or find. If you have more than one prep, you may want to try this for just one class the first year, then develop the other course next year. Teaching with this style will not allow you to have more than two preps and still have your health and sanity. Try and get your administration to understand that this type of teaching will require more time to set up and take down lab equipment than other classes. At the same time, it will significantly increase student engagement and therefore learning. So try to have the school limit the preps you teach to one or two; three gets to be a breaking point. I have done it with more preps, but it was a lot of work, and there was not a lot of personal time for myself and my family; that took a toll on my health. There will be times in the year if you have two preps that you will have to put some extra time in because those messy labs sometimes back up to each other. Once you get in a routine, you will see where those places are for you in your curriculum, and you can plan for them in the future; it does get easier.

Preparation

Set up your equipment the day before, and put it away the same day you finish the lab. This way, you are not scurrying around at the beginning of the day that you need the equipment, and you can inspect and make sure you have everything you need for all of your classes. Early preparation sets you in a better mood for the day you are using the equipment because you know you are prepared. Otherwise, when you are flustered about getting your equipment together the same day as the class, the slightest thing will set you off, and evil thoughts can pile up quickly, limiting your patience. Put the equipment away that day, so others who need it (on your team) can find it for themselves; this will help build relationships with your team if they can count on you to do your part, and they do not have to come behind you and pick up after you. I hate working with those teachers who put nothing back, leaving things dirty for more work for me. Make sure you put in a bit of extra effort to not cause more work for others; this will help build and maintain relationships with colleagues.

I like making my copies a unit at a time, and never the day before or the morning of when I need them. Copy machines have a way of breaking down or being busy when they are needed most. Make your copies, or send in your documents to the copy person, well before needed. You may need to come in early or stay late to take care of this, so it is not a burden to you if the copier goes down. Keep in mind that the copy ladies love those who get their work in early. Be extra nice to the copy lady; she is a busy person, and you may need a favor one day. Treat her with extra special love and give presents occasionally; this will go a long way. You may want to investigate how to go paperless if this is something you struggle with in any way.

Technology will go down at some point. Make sure that you check all of the technology works the way the students will be using it. Get technologically savvy with how to solve the little things that can go wrong. Just like all of the other equipment you will have your students use, you will need to anticipate problems that the students may have. Some websites do disappear or change their format. So if it has been a whole year since you looked at a website you are about to use, go back and look at it before the students do to make sure it is still there and set up the same way. When working with probe-ware, make sure you set it up and go through the lab; to make sure you can look at all data and graphs you want the students to look at, making sure the probes can measure what you want them to. Some probes are not very reliable others are very precise. Find out which probes you will be able to use, which you may want to stay away from, and which you may want to change how you use it, for example, changing a lab into a demo because the data students may get themselves would not come out properly because of some mistake that is commonly made when using that equipment.

Tub your Equipment

If they fit, you can put equipment students will use for the labs in tubs, so they do not play with the equipment before labs. Just putting equipment in tubs, many times, will prevent students from toying with it. Most of the time, they will not touch it until it is time to do the lab. I like to do multiple labs in one class, and it is easy to teach the students to pick the equipment that you want them to use and leave the others alone in the tub, but you do need to take the time to teach this skill early. It is also easy to move the equipment to tables if you have multiple preps during the day or have students who can't keep their hands off the equipment until you do labs. You can also have tubs at the table for other things like pencils, erasers, and calculators, which students may use every day.

Students Who Don't Want to Do Anything

These students do not mind lectures because they do not have to do anything during a lecture. They do not have to think at a very high level when taking tests, and copying is easy to do. Because they are forced to observe and learn when doing investigations, they may tell you they cannot learn this way, which is not true. They can; they do not want to put in the effort to think; they want someone to do it for them. Some may not have ever had to think this way before, and it will be scary for them. Assure them that they do not have to work on anything on their own, but they need to contribute to their team. Try to focus on instilling a growth mindset with these students. They can be quite stubborn at first, but they will eventually buy in once they see the change in themselves bring success. Also, have them focus on the learning, not getting the final answer or product; this will reduce these reluctant students' procrastination. New classes may struggle at first, depending on the makeup of the students and their background. You may have to model how to think for them through Socratic questioning. If you model the thinking too much, they will rely on you or the class telling them the answers at the end and not developing those thinking skills to get the answers themselves. They are just waiting, many times, for you to tell them the answer. As they build their skills with their successes, I give them more independence, and they get better at this mode of thinking and more confident. Believe it or not, many students will thank you for it later. The sooner you let them observe on their own to figure things out, the sooner they will start trying and be resigned to start thinking like a scientist on their own and start enjoying the class. It may take some prepping and promoting on your part to convince them, especially if they have never had to think on their own before. All the beginning of all classes, I feel more like a motivational speaker than a science teacher. If you have a particularly challenging student, you may also want to check with other teachers and find out what works with that student in their class.

While students are learning through investigations, they seem like they are learning like the Karate Kid being trained by Mr. Miyagi. The Karate Kid has no clue that he is actually learning Karate while painting the fence, sanding the floor, putting the wax on, and taking the wax off. Mr. Miyagi was specific regarding how he wanted Daniel to perform the tasks he gave to him repeatedly. You are going to provide activities where they will learn content while at the same time specific skills. Much of the time, they will not know they are actually learning and doing science skills in the activities. Just like the Karate Kid, your students will get frustrated if you do not point out the skills they have learned from time to time. Sometimes you will need to point out how the specific skills and the content they are learning are in context with the other concepts. Like I tell my students regularly, I am trying to teach them common sense, which is not very common, and at the same time have them learn some Biology, Chemistry, or Physics. Common sense is the ultimate goal, getting students to be able to think for themselves.

Be a Good Role Model

While coaching, I was once told that a team takes on the personality of their coach. I believe that can be true for your classes. People do a lot of learning through imitation. So be a good role model for your students and the teachers around you. You are always being watched. If your students see you with a bad attitude or give up, they will too. Think of it like this, whatever you want your kids to do, you better be doing it yourself. "Do as I say, not as I do..." does not happen anywhere successfully. Our brains are made to mimic what we see. Why else would you see boys walking around with the top of their pants around their knees? It cannot be comfortable, and they sure spend a lot of effort keeping them there and not tripping over them (when they do go down, it sure is funny to watch). They do it for no other reason than because those who they look up to are doing the same thing. So be a positive role model; you may not see them mimic you now, but because they saw it, and if they saw it enough, they may just imitate you later when you are not around. When students decide to be more productive, they will draw on actions they saw others do. Students will do and say things they never thought they would, just because you did it in front of them. I know I did. You never know who will be the future leaders. Be sure you emulate characteristics that you would like to see in our leaders, so those characteristics have a better chance of making it into our future leaders (your students). Remember that we, as science teachers, not only want to give students that elusive common sense, but we want them to grow up to be great human beings. We will ensure there are a lot more if we make our future leaders into great human beings; this means we need to be good models of the types of behavior we want our students to have. Those people under our future leaders that we helped make; will emulate them just like our students emulated us.

Final Thoughts

You should ask yourself, "Why should I change how I teach science now and start teaching through investigations?" Instead of being the primary resource for science content in your class by spending your time telling it to your students, you can monitor their engagement and behavior and let the investigations deliver the content for you; this will stimulate their curiosity and allow them to absorb more information and skills more efficiently. When they develop their questions for you, their brains will be ready to accept what you have to say. What we have been doing by giving lectures, quizzes, worksheets, and vocabulary definitions has not been working. Other countries are outscoring us on standardized tests and out-competing us for high-paying jobs here in the United States. Today's students and adults are hard to keep engaged if they are not actually doing something. Teaching through investigations will support our students' engagement. We all learn best by doing and teaching. So this will raise the cognitive level to which all students can discover the science content.

Because this teaching style is naturally differentiated, students naturally will work to their particular level allowing them more success. They will also be independent learners, so they will be able to think for themselves. This will improve our students' capacity in other subject areas, which will improve all standardized test scores for all subject areas. People from different countries are out-competing us in areas that require higher knowledge and technological skills. Teaching through investigations will get students prepared for jobs in the 21st century; this will be done by teaching students skills they will need to use for the rest of their lives. Because of this, our students will be more marketable and be able to compete for those higher-paying jobs that require innovative thinking and technical skills. If all science teachers start teaching this way, we will have more people with common sense and a higher quality workforce. Having more high-quality workers will allow us to stop outsourcing jobs to immigrants from other countries or sending work overseas, thus, putting more Americans to work with higher pay. Through investigations, teaching will improve science literacy nationwide because our students will be doing science instead of just talking about it. The only way students will understand what science is and how it works will be to have them repeatedly do science. Why would we want to rob our children of acquiring the skills and knowledge they will need to have a prosperous future? Switch now and watch the changes that take place in your students, no matter what level they are, and see how this improves our classrooms today.

References

Ba, Harouna and Bram Duchovnay (2002). Science Education and Urban Youth: *A Look at the JASON Project in Philadelphia*. Retrieved July 29, 2008, from www.techlearning.com/db_area/archives/WCE/archives/harouna2.php

Bonwell, C. and J. Edison (1991). Active Learning: *Creating Excitement in the Classroom,* AEHE-ERIC Higher Education Report No.1.Washington, C. Jossey-Bass. ISBN 1-87838-00-87.

Bunday, Karl M., 2006. *Age Segregation in School,* Learn in Freedom.

Campos, Jaclyn & Angela Barton (2004). *Talk in an Urban Elementary Science Classroom: Exploring the Different Perspectives*. Retrieved March 10, 2008, from Columbia University, Urban Science Education Center Department of Math, Science & Technology Teachers College Web site: http://www.tc.edu/centers/urbanscience/papers/Teacher%20and%20Student%20Talk.htm

College Board, 2008, AP Biology Course and Exam Description Effective Fall 2012. The College Board.

Coordinator for Gifted and Talented, K-12 in CFISD (2007, October). *Understanding the Cy-Fair Horizons Program Fall 2007*. In-service Cypress, TX.

Cotton, Kathleen, 1993. *Nongraded Primary Education,* School Improvement Series. Portland, OR: Northwest Regional Educational Laboratory.

Cypress-Fairbanks ISD. 2007-2008 source of District Goals. Retrieved January 13, 2008, from http://www.cfisd.net/aboutour/dgoals.pdf

Cypress Fairbanks ISD Annual Action Plan for 2007-2008 (2007). Retrieved March 24, 2008, from www.cfisd.org.

ExploreLearning (2008). Retrieved 5/2/2008 from www.explorelearning.com/index.cfm?method=cCorp.dspLearnMore.

Ferguson, Ronald F. Ph.D. (2002, December). *Addressing Racial Disparities in High Achieving Suburban Schools*. NCREL Policy Issues, Issue 13.

Garriga, Maria (2007, August 9). *Analysis Contends State Achievement Gap Remains*. New Haven Register.

Golba, Amy (1998). *How does Education in Urban Schools Compare to Suburban Schools?* Anansi's Gourd: Indiana University South Bend Undergraduate Research Journal.

Grossen, Bonnie Ph.D., 1996. *How Should We Group to Achieve Excellence With Equality?* The University of Oregon.

Hambrick, Arlene Ph.D. (2004).*Remembering the Child: On Equality and Inclusion in Mathematics and Science Classrooms.* Retrieved March 10, 2008, from www.ncrel.org/sdrs/areas/issues/content/cntareas/math/ma800.htm

Hayes, Renee (2000). *Tracking and Minority Groups: Silenced Injustice.* The University of Delaware. Retrieved March 14, 2008, from http://ematusov.soe.edel.edu/final.paper.pub/ final2/00000013.htm

Heddens, James W. (1997). *Improving Mathematics Teaching by Using Manipulative.* Back to Edumath 4 (6/97) 50 47.

Herreid, Clyde Freeman (1994). *Case Studies in Science a Novel Method of Science Education.* Journal of College Science Teaching (pp.221-229) NATA Publications Arlington VA.

Johnson, Debra and Fox, Cheryl, 1998. *Critical Issue: Enhancing Learning Through Multiage Grouping,* North Central Regional Educational Library.

Joseph, Pamela Bolotin, Stephanie Luster Bravemann, Mark A Windschitl, Edward R. Mikel, and Nancy Stewart Green (2000). *Cultures of Curriculum.* Mahwah, NJ: Lawrence Erlbaum Associates, Inc.

Katz, Lilian G., 1995. *The Benefits of Mixed-Age Grouping,* EDO-PS-95-8.

McClellan, Diane E. and Kinsey, Susan J., 1999. *Children's Social Behavior in Relation to Participation in Mixed-Age or Same-Age Classrooms,* Early Childhood Research and Practice 1(1).

Melber, Leah (2004). Inquiry for everyone: *Authentic science experiences for students with special needs.* Teaching Exceptional Children Plus, Volume 1, Issue2, Los Angeles, CA.

Mentoring Minds.com (2007). Retrieved 2/2/2008 from http://www.mentoringminds.com/research/researchTAKSPractice.html

Moen, Bonnie R., 1999. *Multi-age Education—Time for a Change,* Paper for Ed. 702 – Psychological Foundations of Education.

National Research Council, 2012 *A Framework for K-12 Science Education: Practices, Crosscutting Concepts, and Core Ideas, The* National Academies Press.

National Science Teachers Association (2003, March 20). NSTA Launches City Science. *NSTA Web News Digest*. Retrieved 3/10/2008 from www.3.nsta.org/main/news/stories/nsta_story.php?news_story_ID=48062

National Science Teachers Association (2008). *The use of Computers in Science Education*. Retrieved 5/2/2008 from www.nsta.org/about/positions/computers.aspx.

Parkinson, Justin (November 22, 2004). *How Computers Can Help in the Class*. BBC News. Retrieved 5/2/2008 from http://newsvote.bbc.co.uk.

Reeder, Eeva (2008). *Measuring What Counts: Memorization Versus Understanding*, Edutopia, The George Lucas Education Foundation. Retrieved 2/2/2008 from http://www.edutopia.org/measuring-what-counts-memorization-versus-understanding.

Reese, Debbie, 1998, *Mixed-Age Grouping: What Does the Research Say, and How Can Parents Use This Information?* Kid Source Online.

Ricks, Irelene (2004). The 50th Anniversary of Brown v. Board of Education: Continued Impacts on Science Education. *Cell Biology Education* 3(3), 146-149.

Salend, Spencer J. (2008). *Creating Inclusive Classrooms: effective and Reflective Practices Sixth Edition*. Pearson Merrill Prentice Hall, Upper Saddle River, NJ, and Columbus, OH.

Schafersman, Steven D. (1991). *An Introduction to Critical Thinking*. Retrieved 2/2/2008 from http://www.freeinquiry.com/critical-thinking.html.

Scherer, Marge (2008, February). *Perspectives / The Thinking Teacher*. Educational Leadership, 65(5), 7-7.

Somerby, Bob (2005, November 21). *Gush, Gush, Sweet Charlotte (Part 4)! When PBS Reviews Charlotte-Meck.* The Daily Howler.

Songer, N.B., H.S. Lee, & R. Kam (2002, February). *Technology-Rich Inquiry Science in Urban Classrooms: What are the barriers to inquiry pedagogy?* Journal of Research in Science Teaching, *39(2),* 128-150.

Sutman, Francis X. (1993). *Teaching Science Effectively to Limited English Students*. ERIC Clearinghouse on Urban Education Digest, Number 87, New York, NY.

Tal, T., R. Geier, & L. Krajcik (2000, April). *Urban Students' Beliefs About Science in Inquiry-based Classrooms*. Paper presented at AERA Conference, New Orleans, LA.

Tayyari, Farnoosh (2004, September). *The Genetic Basis of Intelligence.* The Science Creative Quarterly, Issue 3.

Texas Education Agency. 2009 source of TEKS Chapter 112 Subchapter C. Retrieved December 21, 2017, from http://ritter.tea.state.tx.us/rules/tac/chapter112/ch112c.html#11

Texas Education Agency, District and Schools' Accountability Tables (2006). Retrieved September 29, 2007, from http://www.tea.state.tx.us.html

The Critical Thinking Community (2004). *The Role of Socratic Questioning in Thinking, Teaching, and Learning*. Retrieved 2/2/2008 from http://www.criticalthinking.org/resources/articles/the-role-socraticquestioning-ttl.shtml. Retrieved 2/2/2008.

Tobin, Kenneth & Cristobal Carambo (2004, April). *Expanding the Transformative Potential of Science Education for Inner City Youth*. Paper presented at the annual meeting of the American Educational Research Association, San Diego, CA.

Urban Habitats (2008). *Bringing the Urban Environment into the Urban* 3/10/2008 from www.urbanhabitats.org/v03n01/classroom_app_a.html

Walker, Mark (2008), *The National Science Education Standards*. Retrieved 2/2/2008 from http://www.geocities.com/inquiryscience/nationalscienceeducationstandatds.html.

Wikipedia (2008). Retrieved 2/2/2008 from http://en.wikipedia.org/wiki/Active_learninghttp://www.tc.edu/centers/urbanscience/papers/Teacher%20and%20Student%20Talk.htm

Wikipedia, *Genetics of Intelligence* (2007). Retrieved September 29, 2007 from http//:www.en.wikipedia.org.html

Zagorsky, Jay L. (2006, February). *Do you have to be smart to be rich? The impact of IQ on wealth income and financial distress.* Science Direct.

www.ingramcontent.com/pod-product-compliance
Lightning Source LLC
Chambersburg PA
CBHW031537210526
45464CB00003B/1055